11/20

Introduction to EARTH'S RESOURCES

HOW WE USE
WATER

Nancy Dickmann

Crabtree Publishing Company
www.crabtreebooks.com

Crabtree Publishing Company
www.crabtreebooks.com

Author: Nancy Dickmann
Editorial Director: Kathy Middleton
Editor: Ellen Rodger
Picture Manager: Sophie Mortimer
Design Manager: Keith Davis
Children's Publisher: Anne O'Daly
Proofreader: Debbie Greenberg
**Production coordinator and
  Prepress technician:** Ken Wright
Print coordinator: Katherine Berti

Photographs (t=top, b=bottom, l=left, r=right, c=center)
Front Cover: All images Shutterstock
Interior: Alamy: 21; iStock: Antonistock 4, Bridgendboy 6,
George Clerk 12, Jonathan Austin Daniels 5, fivepointsix
26, Kaisersosa67 10, Dani Kancil 16, nolimitpictures 19,
Nordic Moonlight 13, Patty C 27, Franz Rido7, Pete Saloutos
24, tonda 15, zu-09 14; Shutterstock: Africa Studio 29cl,
Andrea Danti 23 Grigvovan 28t, Dmitriy Gumenyuk 1,
JasrolavaV 8–9t, Vasiliy Merkushev 8–9b, NavinTar 18,
Andrey Popov 28b, Catherine L Prod 25. Saowakin Sukrak
17, Lullia Timofeeva 22, Vasca 11, Zerbor 29br; US Federal
Government: United States Coastguard 20.
All facts, statistics, web addresses and URLs in this book
were verified as valid and accurate at time of writing. No
responsibility for any changes to external websites or
references can be accepted by either the author or publisher.

Library and Achives Canada Cataloguing in Publication

Title: How we use water / Nancy Dickmann.
Names: Dickmann, Nancy, author.
Description: Series statement: Introduction to Earth's resources |
  Includes bibliographical references and index.
Identifiers: Canadiana (print) 20200284479 |
  Canadiana (ebook) 20200284509 |
  ISBN 9780778781981 (softcover) |
  ISBN 9780778781844 (hardcover) |
  ISBN 9781427126023 (HTML)
Subjects: LCSH: Water use—Juvenile literature.
Classification: LCC TD348 .D53 2020 | DDC j333.91/13—dc23

Library of Congress Cataloging-in-Publication Data

Names: Dickmann, Nancy, author.
Title: How we use water / Nancy Dickmann.
Description: New York, NY : Crabtree Publishing Company, 2021. |
  Series: Introduction to earth's resources | Includes index.
Identifiers: LCCN 2020029717 (print) | LCCN 2020029718 (ebook)
  ISBN 9780778781844 (hardcover) |
  ISBN 9780778781981 (paperback) |
  ISBN 9781427126023 (ebook)
Subjects: LCSH: Water use--Juvenile literature.
Classification: LCC TD348 .D53 2021 (print) | LCC TD348 (ebook) |
  DDC 333.91/13--dc23
LC record available at https://lccn.loc.gov/2020029717
LC ebook record available at https://lccn.loc.gov/2020029718

Crabtree Publishing Company
www.crabtreebooks.com   1-800-387-7650
Published in 2021 by Crabtree Publishing Company

Copyright © Brown Bear Books Ltd 2020

**Published in Canada
Crabtree Publishing**
616 Welland Ave.
St. Catharines, ON
L2M 5V6

**Published in the United States
Crabtree Publishing**
347 Fifth Ave
Suite 1402-145
New York, NY 10016

Printed in the U.S.A./082020/CG20200710

In Canada: We acknowledge the financial support of the
Government of Canada through the Canada Book Fund for
our publishing activities.

Contents

What Is Water?

Water is everywhere: in the air, covering the land, and in our homes. We use it every day.

When we turn on the faucet, the water that pours out is a liquid. But water can change. At temperatures below 32 °F (0 °C), water freezes into a solid called ice. Above 212 °F (100 °C), it boils and turns into a **gas** called **water vapor**.

Steam is another name for water in its gas form.

At the top of Mount Everest, water boils at 156.2 °F (69 °C).

When water vapor in the air hits the cold surface of a glass of ice water, it cools to form droplets on the outside.

Three Forms

Water exists on Earth in all three of its forms. There is liquid water in oceans, rivers, and lakes. Rain falls from the sky. Ice often forms in winter, and there is ice all year at the North and South poles. There is water vapor in the air around us and in clouds in the sky.

Salty ocean water freezes at about 28.4 °F (-2 °C).

Why Is Water Important?

All living things, from tiny ants to tall trees and giant whales, need water to survive.

Earth is the only planet in our solar system with liquid water on the surface. It is also the only planet where we know that life exists. Earth's water allows all types of living things to grow and thrive.

Different body parts have different amounts of water. Muscles contain a lot of water, and even bones have some!

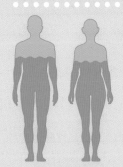

Water makes up about **60%** of the average human body.

A jellyfish's body is about **95%** water.

Our bodies lose water when we sweat and breathe. You should drink plenty of water to replace it.

Using Water

Water helps humans and other animals to control their body temperature by sweating. Water also helps carry **nutrients** around the body. It carries away waste products when we use the toilet, and it keeps our joints such as knees and hips working. Water also protects delicate body parts such as the brain.

The Water Cycle

Earth's water is always moving. It moves between Earth, the sky, and the oceans. This is the water cycle.

Water falls as rain or snow. It soaks into the ground and runs into rivers, lakes, and oceans. The Sun's heat makes water **evaporate**, or turn into vapor.

Clouds and Rain

Water vapor floats up into the air and **condenses**, or turns into tiny droplets. The droplets make clouds. The water droplets fall back to Earth as rain or snow. Then the cycle starts again.

Two-thirds of Earth is covered with water.

The Water Cycle

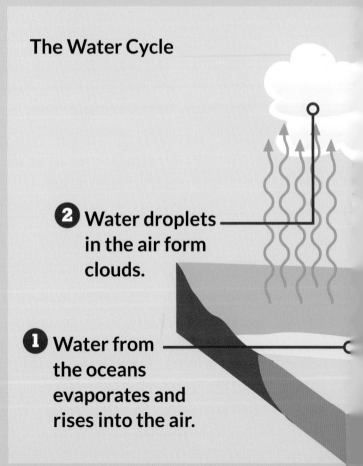

2 Water droplets in the air form clouds.

1 Water from the oceans evaporates and rises into the air.

Everything on Earth needs water to survive.

97% of Earth's water is held in the oceans, and **2%** is frozen in polar ice caps. That leaves just **1%** for people to use!

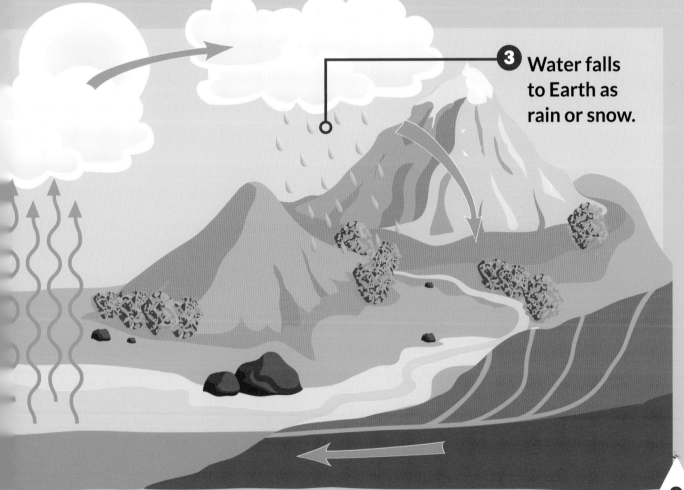

3 Water falls to Earth as rain or snow.

How Do We Use Water?

At home, we take a constant supply of fresh water for granted. But how do we actually use it?

On an average day, you probably brush your teeth, take a shower, and flush the toilet several times. But what about cooking and drinking? How often does someone in your house run the washing machine or the dishwasher? Are there plants that need watering?

The different ways that we use water at home can really add up.

Watering grass lawns uses a lot of water.

The average U.S. family uses about **300 gallons** (1,000 l) of water per day.

Running a bathroom faucet uses about **2 gallons** (7.6 l) of water per minute.

Flushing the toilet uses **26.7%** of a home's water.

There are often separate pipes for hot and cold water.

Coming and Going

Most homes have two sets of pipes. One set brings in the fresh, clean water that comes out of our faucets. These pipes also supply dishwashers and other appliances. The other set of pipes carries away dirty water to be cleaned. It collects waste water from sinks, toilets, bathtubs, and showers.

Water and Industry

Water is not just used in the home. Factories, farms, and businesses use water, too.

Farming is one of the biggest uses of water. Farmers need water for their animals to drink. They rely on rain to keep their **crops** healthy, but they often must water them too. This can use a huge amount of water. In total, farming uses more water than any other industry.

Watering crops is called irrigation. It lets farmers grow crops in dry places.

About **70%** of the water used around the world is for watering crops.

Water in Factories

Many of the products we use every day are made in buildings called factories. Water is an ingredient in many products. It is also used for cooling factory machinery, washing materials, and changing them into other forms. Some industries, such as makers of paper, use a lot of water.

In the last **40 years**, worldwide water use has gone up by **33%**.

At a paper mill, water is used to break down wood to form paper.

Not Enough Water

Living things all depend on water. But what happens when there is not enough?

When an area gets much less rain than usual over a long period of time, it is called a **drought**. Droughts can last for months or even years. Without enough water, crops and other plants may die. That means there is not enough food for the people and animals that depend on them.

Rivers and lakes can dry up during a long drought.

In just **one year** in the United States, drought destroyed crops worth **$17 billion**.

A cactus stores water in its thick, fleshy stem.

Deserts

Some places, such as **deserts**, get very little rain. But this is not a drought. It is the normal climate pattern for those areas. Desert plants and animals are **adapted** to the dry conditions. They are able to find and store water and can get by with very little of it.

About **33%** of all the land on Earth is desert.

Too Much Water

Not having enough water is a serious problem, but having too much can be just as dangerous.

Floods happen when a large amount of water overflows from oceans, rivers, lakes, or underground and spreads onto the land. Heavy rainfall can make rivers overflow their banks. Hurricanes and other storms can also cause floods by pushing water from the ocean onto the land.

Rescue workers use boats to help people escape floodwater.

Around the world, floods cause nearly **$40 billion** worth of damage a year.

When the Mississippi River flooded in 1927, water covered **16 million acres** (6 million hectares) of land.

A powerful flash flood can sweep away anything in its path.

Flash Floods

With some floods, the waters rise slowly over a period of days or weeks. People have time to get away. Other floods happen very quickly. They are called flash floods. The rain falls so fast that the ground cannot soak it up. Flash floods are very dangerous and can turn roads into fast-flowing rivers.

Clean Water

We need clean water for drinking and washing. Using dirty water can cause disease.

Fresh drinking water usually comes from a lake, river, or well. It is cleaned at a treatment plant before going into homes. First, large **particles** are allowed to sink to the bottom. Then the water is sent through sand to catch any smaller particles. Chemicals or **radiation** is used to kill any **germs** in the water.

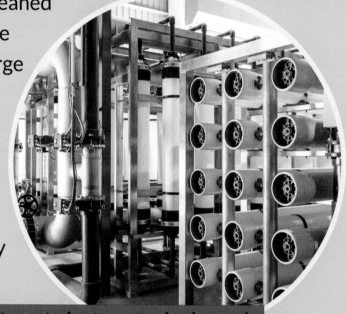

At a treatment plant, water is cleaned and tested to make sure it is safe.

There are about 170,000 public drinking water treatment systems in the United States.

Water Disease

Many people around the world don't have access to clean water. They must use water that often contains harmful germs. It might be **contaminated** with human or animal waste. Drinking dirty water can cause diarrhea, vomiting, and serious illness.

Illnesses caused by dirty water kill **1.6 million** young children each year.

Boiling water for a couple of minutes should kill any germs in it.

Polluted Water

When water gets very dirty, we say it is polluted.

Water gets **polluted** when chemicals run into rivers and streams. The chemicals might come from factories. They might be washed into rivers and streams from farmland. They can even come from household cleaning products going down the drain.

In 2010, the Deepwater Horizon oil rig exploded in the Gulf of Mexico. It caused the biggest oil spill in U.S. history.

More than **8,000 animals** (sea turtles, birds, and mammals) died in the six months after the spill.

The explosion spilled up to **39 million gallons** (148 million l) of oil.

BP, the company that owned the oil rig, had to pay **$40 billion** in fines and cleanup costs.

The explosion on the Deepwater Horizon caused a huge fire.

Volunteers carefully clean an oil-coated bird.

Oil Spills

Oil spills are another type of pollution. Oil floats on the ocean's surface. That affects sea mammals and birds. Oil coats an animal's fur or feathers. It stops fur from keeping the animal warm and makes feathers less waterproof. Drinking polluted water of any type can kill plants, animals, and humans.

Electricity from Water

Moving water has a lot of energy. It can be used to produce electricity.

Long ago, people used moving water to turn waterwheels. The turning wheel was used to turn huge stones for grinding grain. Today, we use the movement of water to create a form of power called **hydroelectricity**. This kind of power makes up 16% of the world's electricity supply.

The Itaipu Dam in South America is one of the world's largest hydroelectric plants.

Building the dam took **18 years** and cost **$27 million**.

The construction used enough concrete to build **210 stadiums**.

The dam can produce enough electricity for **30 million** people.

Hydroelectricity is a form of renewable energy that does not cause pollution.

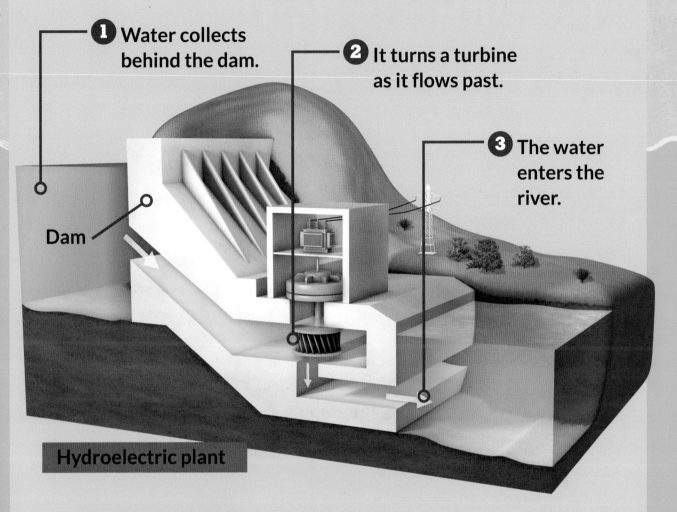

1 Water collects behind the dam.

2 It turns a turbine as it flows past.

3 The water enters the river.

Dam

Hydroelectric plant

How it Works

Many hydroelectric plants are built on rivers.
A large strong **dam** is built to block the water.
The water builds up behind the dam. It is allowed
to travel down a narrow passage, where it turns
the blades of a fan-like **turbine**. A **generator**
changes the spinning motion into electricity.

Water for Fun

Water is useful. In fact, it's essential to life. But it can also be fun!

Humans have been swimming for thousands of years, although it has only become widespread as recreation in the last 200 years. Knowing how to swim can save your life, and it is also a great way to exercise and have fun. Many swimmers take part in races or in other events, such as diving or water polo.

Some people compete in synchronized swimming, which is like dancing in the water.

The world's fastest swimmers can swim 165 feet (50 meters) in under 21 seconds.

Enjoying the Water

People have come up with many different ways to enjoy water, from bodyboarding and surfing to waterskiing. There are also water parks where visitors can enjoy splashing and zooming down waterslides. Being wet is fun, and it can also help your body cool down on a hot day.

The world's longest waterslide is 8,933 feet (2,723 m) long.

Adding water to a slide makes for a faster, more slippery ride.

Saving Water

Most of the water on Earth is saltwater. People need freshwater for drinking.

The city of Cape Town in South Africa was hit by a long drought. By 2018, they were about to run out of water. The government asked people to reduce the amount they used. They banned washing cars and filling swimming pools. They fined people, or charged them money, if they used too much water. It worked!

In Cape Town, people sometimes lined up to fill water containers.

During the crisis, residents were limited to using **13.2 gallons** (50 l) per day.

By the end, the city was using **15.5%** less water per day.

A rain barrel collects rainwater so that it can be used on the garden.

Reducing and Recycling

There are two main ways to save water. The first is to use less, such as by turning off the faucet while you brush your teeth. Another way is to reuse water. Rainwater can be used to water plants. Bathwater can be used to flush toilets.

What Can I Do?

Here are some tips on how you can save water and help keep it clean.

- Take short showers instead of baths.

- Don't cut your grass too short. Longer grass holds water and will need less watering.

- Only run the dishwasher or washing machine when it is full.

- Peel and wash vegetables in a large bowl of water instead of under running water.

- Never flush wipes, medicine, or trash down the toilet. They can cause pollution.

Quiz

How much have you learned about water? It's time to test your knowledge!

1. What are two names for water in its gas form?

a. ice and steam
b. milk and glue
c. water vapor and steam

2. What is the water cycle?

a. a bicycle that's really hard to ride
b. the process that takes water from Earth's surface into the air and then back down again
c. a way of recycling water in a factory

3. What happens at a hydroelectric plant?

a. water is cleaned and recycled
b. the energy of moving water is turned into electricity
c. new water-saving gadgets are invented

4. What is a long period with less rain than usual called?

a. summer
b. desert
c. drought

5. Which industry uses the most water?

a. farming
b. making paper
c. running water parks

Answers on page 32.

Glossary

adapted Having developed behavior or physical features that are suited to a certain environment

condenses Turns from a gas into a liquid

contaminated Something that has been made dirty

crops Plants that are grown for food

dam A barrier built across a river to stop water from flowing

deserts Areas where the normal conditions include very low rainfall

drought A long period with less rain than usual for the area

electricity The flow of current that we use to run gadgets, motors, lights, and more

evaporate To turn from a liquid into a gas

gas A substance that spreads out to fill all the available space

generator A device that changes spinning motion into electricity

germs Tiny living things that can cause illness

hydroelectricity Electricity produced by the movement of water

nutrients Chemicals that a person, animal, or plant needs to grow and stay healthy

particles Very tiny pieces of something

polluted Made dirty by harmful substances

radiation Waves of certain types of energy that can damage living things

renewable Able to be replaced rather than being used up entirely

turbine A device with blades that spin when a gas or liquid flows past it

water vapor Water in the form of a gas

Find out More

Books

Castellano, Peter. *Water Conservation* (Where's the Water?). Gareth Stevens, 2016.

Dickmann, Nancy. *Harnessing Hydroelectric Energy* (Future of Power). PowerKids Press, 2017.

Ganeri, Anita and Chris Oxlade. *Water Cycle* (Geo Detectives). QED Publishing, 2019.

Olien, Rebecca Jean. *Saving Water* (Water in Our World). Capstone Press, 2016.

Websites

www.amnh.org/ology/features/bigideas_water/index.php
Visit this website to find more information about water.

www.epa.gov/watersense/how-we-use-water
This website shows how much water we use for different activities.

www.nationalgeographic.co.uk/environment/2019/04/floods-explained
Find out what makes floods happen and the damage they can cause.

Index

Quiz answers
1. c; 2. b; 3. b; 4. c; 5. a